O
is for Ohio

Written by Kelley Clark
Illustrated by James Balkovek

outskirts
press

To my husband...
For making everything possible and creating OUR book

O is for Ohio, the reason for this book.
So many things you'll learn inside,
once you take a look!

Each letter is a topic
that's special to our state.
Fun facts and bits of history
given to you straight.

So pull up a chair… lie on a bed…
find the place you like to read.
There are also lots of pictures…
go slowly, it's not about speed.

O is for Ohio,
for people any age.
We hope you will enjoy it,
so let's go…turn the page!

A is for the Amish, living simply on Ohio farms. Homemade clothes and horse-drawn buggies are just a couple of their charms.

Holmes County is a special part of Ohio where half the people who live there are Amish. Called "The Gentle People", the Amish value family and hard work. Modern inventions like cars, TV and power tools aren't allowed because they make daily living too easy. Another big difference is that all Amish children go to the same one-room schoolhouse, but only until they complete the eighth grade. Then, the teenage boys and girls work with community elders to learn important skills that will help them earn money as adults.

B is for Buckeye, seeds from Ohio's state tree. Nut-like, shiny brown ones that bring luck to you and me.

Did you know every state has official symbols that stand for its natural treasures and way of life? Ohio's symbols include a state beverage—tomato juice; a state insect—the ladybug; a state bird—the cardinal; and even a state amphibian—the spotted salamander! Ohio's state tree gets its "Buckeye" name from the large brown seeds it produces. Buckeyes are believed to bring good luck, especially to those that carry them in their pockets as charms.

C is for the Cities. Lots begin with the letter "C".
Columbus, Cleveland then Cincinnati are Ohio's biggest three.

Besides being its largest city, Columbus is also Ohio's present day capital but it wasn't always. Chillicothe was the first. Zanesville was the second. The third was … Chillicothe again! It took Ohio until 1812 to settle on a plan to build a new capital in the center of the state named after Christopher Columbus. Home to The Ohio State University, over 875,000 Ohioans now live there making Columbus one of the fastest-growing large cities in the United States.

Cincinnati

Toledo

Lorain

Akron

Canton

Youngstown

Cleveland

Dayton

Columbus

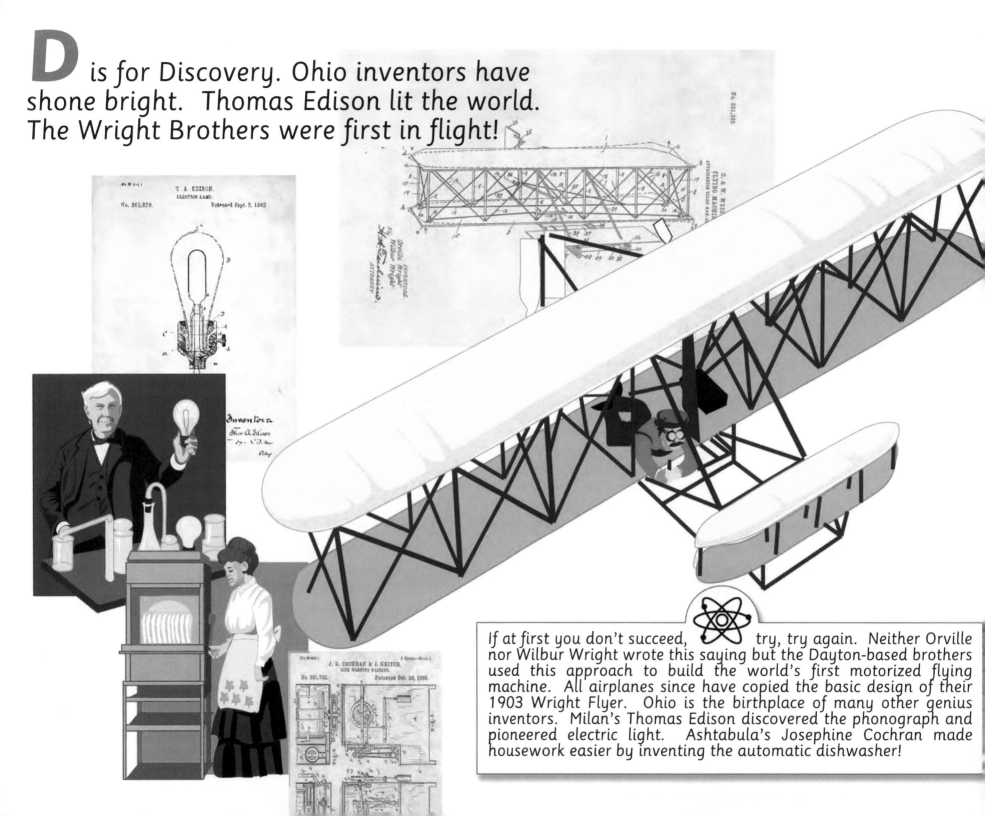

D is for Discovery. Ohio inventors have shone bright. Thomas Edison lit the world. The Wright Brothers were first in flight!

If at first you don't succeed, try, try again. Neither Orville nor Wilbur Wright wrote this saying but the Dayton-based brothers used this approach to build the world's first motorized flying machine. All airplanes since have copied the basic design of their 1903 Wright Flyer. Ohio is the birthplace of many other genius inventors. Milan's Thomas Edison discovered the phonograph and pioneered electric light. Ashtabula's Josephine Cochran made housework easier by inventing the automatic dishwasher!

E is for Lake Erie, on Ohio's north coast.
Across the waves is Canada, between are shipwrecked ghosts.

LAKE ONTARIO

LAKE HURON

CANADA

MICHIGAN

MAPLE SYRUP

NEW YORK

LAKE ERIE

OHIO

PENNSYLVANIA

What do the number 1,400, a sea monster, and Dr. Seuss have in common? Lake Erie, the smallest, shallowest and roughest Great Lake. Home to 1,400 shipwrecks and a sea monster named Bessie, Lake Erie's waters were once so polluted that Dr. Seuss wrote, "I hear things are just as bad up in Lake Erie," in his classic children's book, The Lorax. Years later, he changed the line when enough concerned Ohioans convinced him that Lake Erie's water quality had greatly improved!

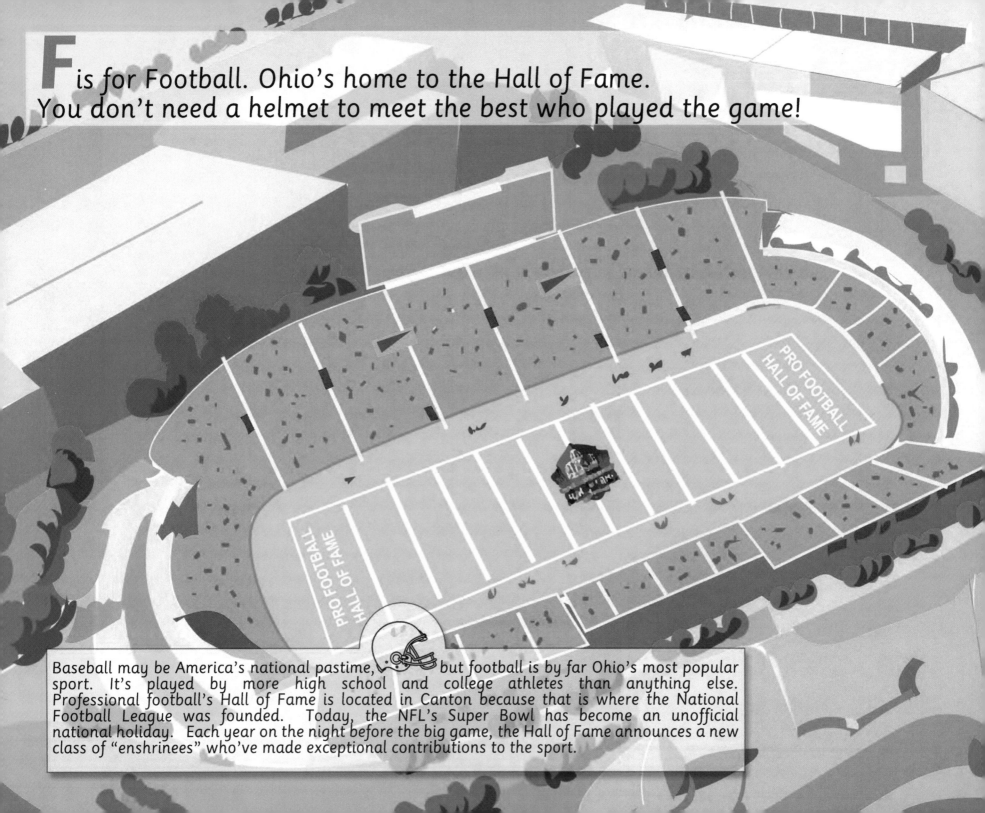

F is for Football. Ohio's home to the Hall of Fame. You don't need a helmet to meet the best who played the game!

Baseball may be America's national pastime, but football is by far Ohio's most popular sport. It's played by more high school and college athletes than anything else. Professional football's Hall of Fame is located in Canton because that is where the National Football League was founded. Today, the NFL's Super Bowl has become an unofficial national holiday. Each year on the night before the big game, the Hall of Fame announces a new class of "enshrinees" who've made exceptional contributions to the sport.

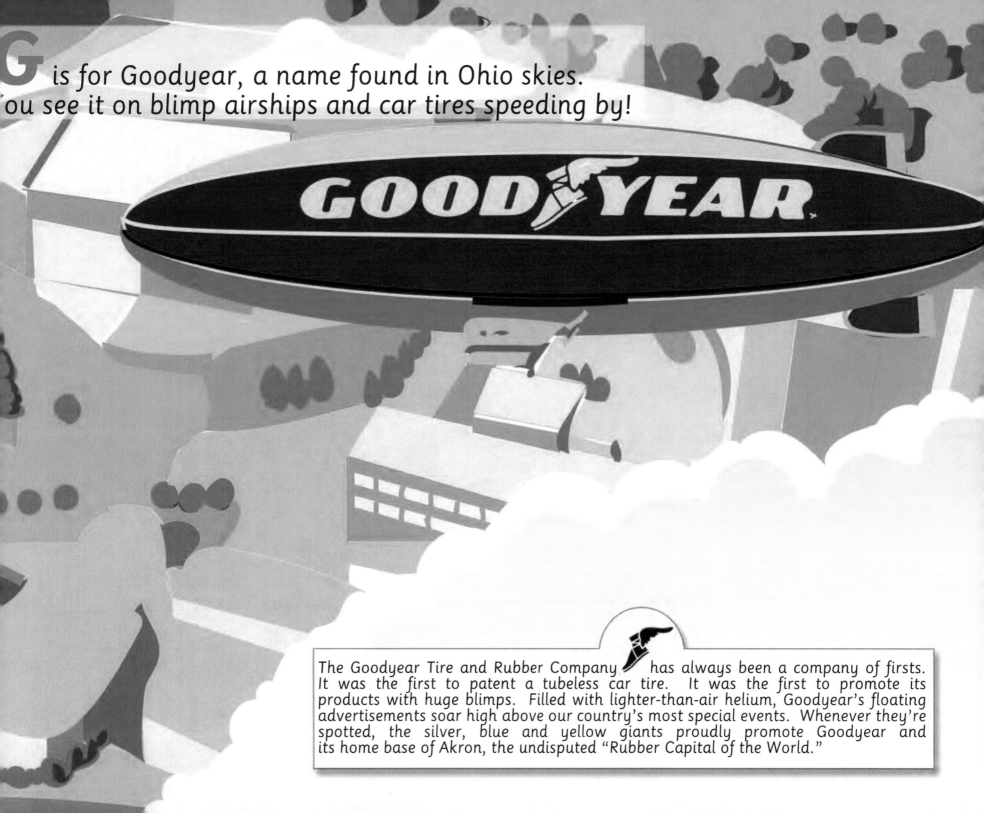

G is for Goodyear, a name found in Ohio skies.
You see it on blimp airships and car tires speeding by!

The Goodyear Tire and Rubber Company has always been a company of firsts. It was the first to patent a tubeless car tire. It was the first to promote its products with huge blimps. Filled with lighter-than-air helium, Goodyear's floating advertisements soar high above our country's most special events. Whenever they're spotted, the silver, blue and yellow giants proudly promote Goodyear and its home base of Akron, the undisputed "Rubber Capital of the World."

H is for Ohio's Hopewell peoples, their mounds have a history.
Shaped like birds and animals, why they made them is a mystery.

Almost 2,000 years ago, Native Americans living in the Ohio River Valley built large earthen mounds shaped like birds, animals and snakes. No one knows for sure why they were made. Artifacts such as jewelry, carvings and tools were often buried inside. Some think these great works served as meeting places for feasts and rites of passage. Others believe they helped the Hopewell track the moon, sun and stars.

I is for Industry. Ohio's one hard-working state.
Steel, rubber ... oil, gas ... plus farming makes us great!

Winter, spring, summer or fall, coal from Ohio's south hills powered steamboats on the rivers. Blast furnaces in the east produced steel that built America's great cities. Food grown on Ohio farms fed the immigrants that came to work in the state's many factories. The products made here helped the United States become the world's biggest economy. Today, agriculture is still the Buckeye State's largest industry. High-tech manufacturing, green energy and medical services developed at Ohio's world-class universities are the jobs of the future.

J is for Jam, or maybe you like jelly.
The best is made in Orrville. Both will please your belly.

Many years ago, a hard-working farmer named Jerome Monroe Smucker learned an important lesson: the sweeter the fruit, the tastier the jam. So, using the sweetest apples grown on trees planted by Johnny Appleseed, "J.M." began selling the tastiest apple butter from the back of his horse-drawn wagon. Over 120 years later, his family still runs the J.M. Smucker Company he founded in tiny Orrville, Ohio. Now, it's one of the world's largest suppliers of jams, jellies, peanut butter, ice cream toppings and more!

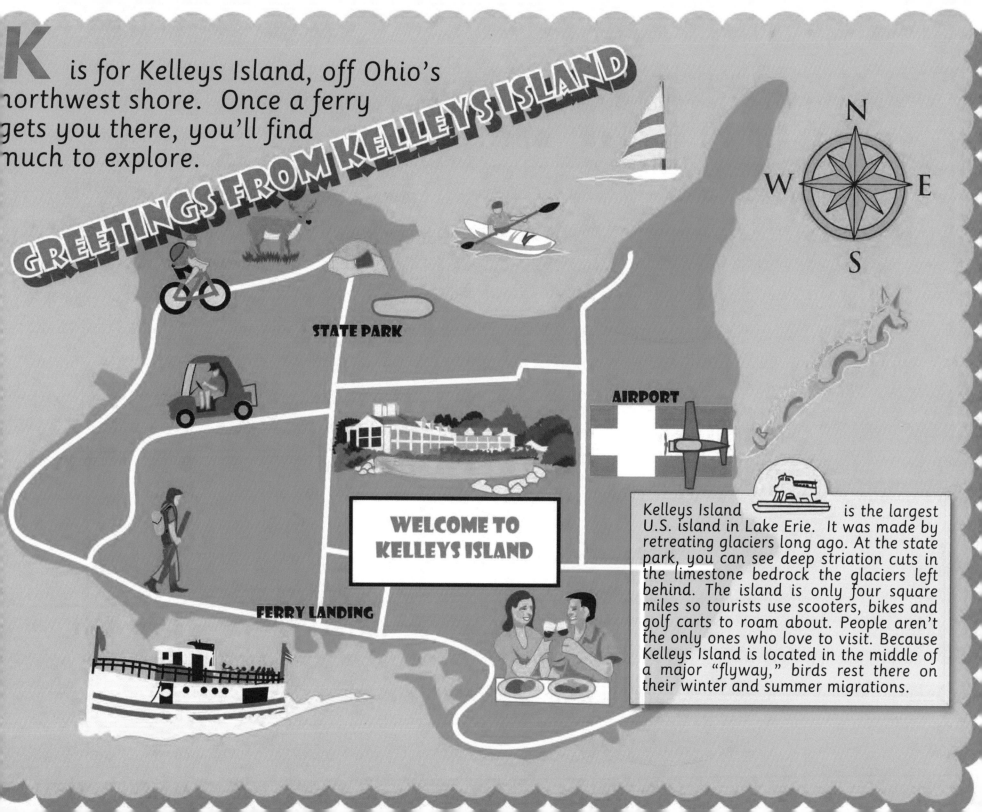

K is for Kelleys Island, off Ohio's northwest shore. Once a ferry gets you there, you'll find much to explore.

GREETINGS FROM KELLEYS ISLAND

STATE PARK

AIRPORT

WELCOME TO KELLEYS ISLAND

FERRY LANDING

Kelleys Island is the largest U.S. island in Lake Erie. It was made by retreating glaciers long ago. At the state park, you can see deep striation cuts in the limestone bedrock the glaciers left behind. The island is only four square miles so tourists use scooters, bikes and golf carts to roam about. People aren't the only ones who love to visit. Because Kelleys Island is located in the middle of a major "flyway," birds rest there on their winter and summer migrations.

L is for the Locks, found along Ohio's canals.
Loaded barges sailed through them, pulled by mules named Sal.

The Ohio and Erie Canal was an important inland waterway that linked the Great Lakes to the Gulf of Mexico by way of the Ohio and Mississippi rivers. Thousands of workers were paid $0.30 a day to dig the trenches and clear the land. When it was done, there were 146 locks that "lifted" the boats along the 308-mile north/south route. The first barge left Akron on July 3, 1827. For the next 30 years, the towns along the canal boomed until it became much cheaper to move goods and people by train.

M is for Mother of Presidents, seven came from Ohio. Two G's, three H's, an M and a T, had it written in their bios.

ULYSSES S. GRANT 1869-1877

RUTHERFORD HAYES 1877-1881

JAMES GARFIELD 1881

BENJAMIN HARRISON 1889-1893

WILLIAM MCKINLEY 1897-1901

WILLIAM TAFT 1909-1913

WARREN HARDING 1921-1923

In terms of birth, Virginia has "produced" eight U.S. presidents, one more than Ohio's seven. But, now the tricky part! Virginia's fifth president, William Henry Harrison, spent his adult life in Ohio, where he was an elected U.S. Senator. William Henry was also the great-grandfather of Ohio's very own fourth president, Benjamin Harrison! The other Buckeye presidents were Ulysses S. Grant, Rutherford B. Hayes, James Garfield, William McKinley, William Howard Taft and Warren G. Harding.

is for NASA Astronauts! Many Ohioans have been to space.
All brave men and women, one even "leapt" on the moon's face.

Maybe it's because Ohio is the birthplace of flight, but no one knows exactly why so many NASA astronauts were born in the Buckeye State. One thing's for sure, though, Ohioans love their space pioneers. Important ones were Cambridge's John Glenn-the first American to orbit earth, Wapakoneta's Neil Armstrong-the first person to ever walk on the moon, Cleveland's Jim Lovell-Apollo 13 commander, and Akron's Judith Resnick-the first Jewish-American in space, who died when the Space Shuttle Challenger was destroyed.

O is for Orchestras. Ohio's are all first class. Many musicians playing together...strings, woodwinds, percussion and brass!

Violins, tubas, and drums—oh my! An orchestra is a large group of musicians all playing the same song, at the same time. How many different musicians and instruments are involved depends on the music being played. A conductor stands up front, holding a tiny baton. With wild hand and arm movements, she unites the players, sets the tempo and shapes the sound. From Toledo to Youngstown, Ohio has seven professional orchestras. Countless more youth orchestras are made up of skillful students throughout the state.

P is for Paczki, a tasty Fat Thursday treat.
Polish Ohioans make them, to celebrate and eat.

Ohio is home to people of all races and cultures. Often, when someone moves to a new country, they bring traditions and foods from their homeland. Paczki donuts are a great example. Pronouned, "pownch-kee", these fluffy, deep-fried dough balls are stuffed with sweet filling and covered with powdered sugar and icing glaze. Polish people throughout Ohio eat their precious Paczki, especially around the Easter Holidays. Sometimes there's even an official Paczki-Day complete with paczki eating contests.

Q is for Queen City, Porkopolis is its other name.
Reds and Bengals fans defend it ... visit Cincinnati to catch a game.

A nickname is a substitute for the real name of a person, place, or thing. Often nicknames convey feelings of love while also poking gentle fun. Cincinnati is blessed with many. In the mid 1800s, it was called Porkopolis because Cincinnati was the country's main hog packing center, and herds of pigs walked the streets. It was also the largest U.S. city not located on the Atlantic Ocean. Because of this, most Cincinnatians liked The Queen of the West, or Queen City nickname better!

R is for Rock and Roll, a CLE deejay started it all.
Driving teenagers wild at his Moondog Coronation Ball!

ELVIS LIVES!

C'mon TUNE IN!

ALAN FREED'S

THE MOON DOG HOUSE

BLUES RHYTHM JAZZ

NIGHTLY at 11:15 PM - 1:00 AM
ON RADIO STATION WJW

HE SPINS 'EM NEED HE'S HEP, THAT FREED

SPONSORED BY

RECORD RENDEZVOUS
· 300 PROSPECT AVE ·

LONG LIVE ROCK

In 1952, when Alan Freed called the new, up-tempo music he was playing on the radio "Rock and Roll," he not only coined a phrase that inspired a generation, he named a lifestyle that forever rocked the world! Besides dealing with typical teenage issues like fast cars and lost boyfriends, early rock and roll songs talked about important social problems that young people of all races were actually facing. Because of his impact, Alan Freed was part of the first group inducted into Cleveland's Rock and Roll Hall of Fame.

S is for Swallowtail, the shape of Ohio's burgee flag.
It stands for history, roads and waterways,
may we never see it sag!

Ohio's unique swallowtail flag is unlike any other state's. Artful and strong, its five red and white stripes represent Ohio's roads and waterways. The bold blue triangle symbolizes Ohio's hills and valleys. The seventeen white stars inside the blue honor the sixteen states Ohio joined when it was admitted to the union. Finally, a proud white "O" surrounds a red disc. These stand for the first letter in "Ohio" and are meant to look like a Buckeye. The flag is so amazing, it took a Boy Scout from Junction City to create the "official" method for folding Ohio's burgee.

that entertained us on our dates.

When you're talented, you 🎂🎂🎂 have a special skill or gift. Applying this meaning to Ohio, there's no denying that the Buckeye State has a real talent for churning out famously talented people. Award-winning actors like Dorothy and Lillian Gish, Paul Newman and Halle Berry. Legendary music-makers like Tracey Chapman, Dean Martin, Devo and John Legend. Game-changing athletes, activists and agitators like Jack Nickalus, Gloria Steinem and John Brown. The list of outstanding Ohioans just keeps growing. Maybe one day, your name will be on it!

U is for Underground Railroad, all over Ohio, stops existed. Safe houses helped slaves escape. Our free state citizens resisted.

Cold rain hits your face but your empty stomach won't let you feel it. Five days, you and your family have been running but you can't shake the hound dogs barking in the distance. Across the river is free state Ohio. So close, but your new baby can't swim. Suddenly, there's a light on the other side. Could it be? Is this your next stop along the secret network of hiding spots that became known as the Underground Railroad? Brave Ohioans played a vital role in its creation. In Cincinnati, Levi Coffin helped three thousand slaves reach freedom in Canada. In Ripley, John Rankin and John Parker used signal lanterns to ferry slaves across the Ohio River from Kentucky where slavery was legal.

V is for great Victories, Ohio teams have won big games!
One involved basketball, and a player named LeBron James!

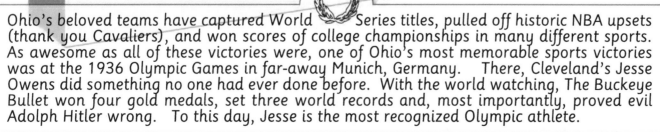

Ohio's beloved teams have captured World Series titles, pulled off historic NBA upsets (thank you Cavaliers), and won scores of college championships in many different sports. As awesome as all of these victories were, one of Ohio's most memorable sports victories was at the 1936 Olympic Games in far-away Munich, Germany. There, Cleveland's Jesse Owens did something no one had ever done before. With the world watching, The Buckeye Bullet won four gold medals, set three world records and, most importantly, proved evil Adolph Hitler wrong. To this day, Jesse is the most recognized Olympic athlete.

is for the Women, who represented Ohio proud. Black, brown, white ... all bold and true ... their voices rang out loud.

Indian chief Sitting Bull called her Little Sure Shot. Buffalo Bill made her his highest paid performer. Who would've thought that tiny "Annie Oakley" (born Phoebe Ann Mosey in a log cabin in Darke County, Ohio) would one day grow up to wow movie stars and royalty worldwide? It didn't matter if it was shooting targets over her shoulder, or knocking dimes out of midair, little Miss Sure Shot always amazed her audiences. More importantly, Annie was a trailblazer who broke down barriers by proving that women are just as capable as men, when given the chance.

X is for Railroad Crossings. Steel rails made Ohio strong. Moving goods and people, hear the whistle and click-clack song.

Ding-Ding-Ding! Flash-Flash-Flash! Over 5,000 miles of railroad tracks and their noisy crossings can be seen all over the Buckeye State. Morning, noon and night, speeding trains take Ohioans and the things they make to distant places. Crops, cars, cows and more. Next time you're stopped at the red flashing lights, have fun trying to count the boxcars rushing past and guessing what's inside.

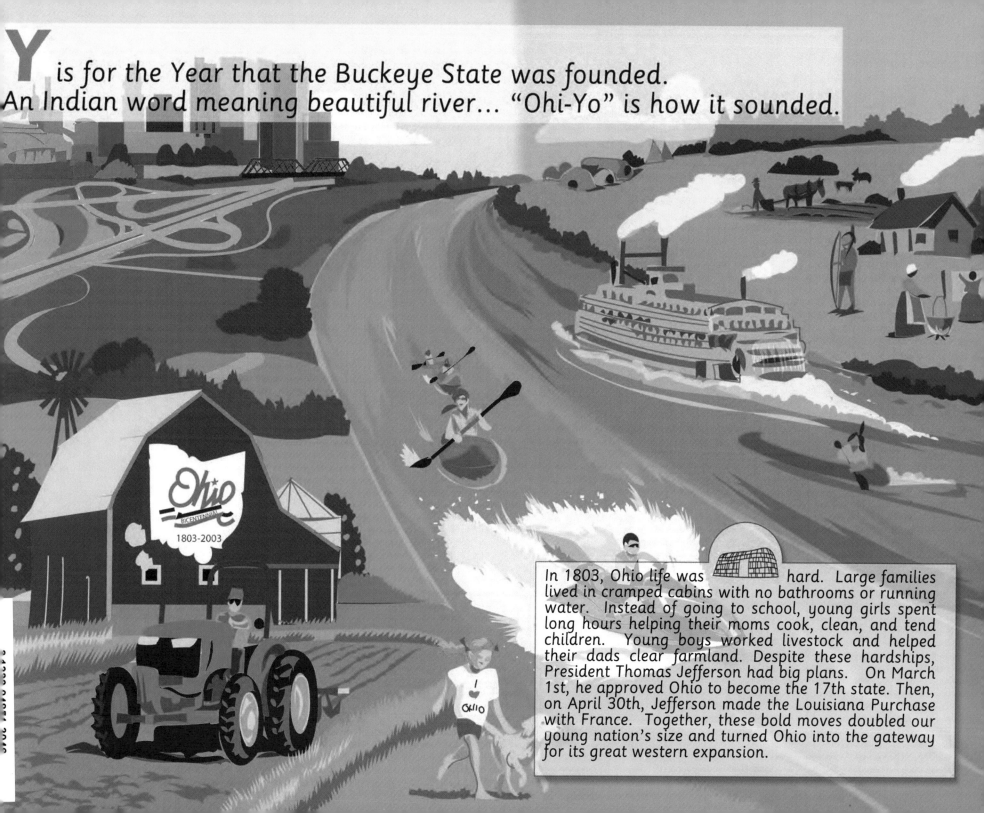

Y is for the Year that the Buckeye State was founded.
An Indian word meaning beautiful river... "Ohi-Yo" is how it sounded.

Ohio
BICENTENNIAL
1803-2003

In 1803, Ohio life was hard. Large families lived in cramped cabins with no bathrooms or running water. Instead of going to school, young girls spent long hours helping their moms cook, clean, and tend children. Young boys worked livestock and helped their dads clear farmland. Despite these hardships, President Thomas Jefferson had big plans. On March 1st, he approved Ohio to become the 17th state. Then, on April 30th, Jefferson made the Louisiana Purchase with France. Together, these bold moves doubled our young nation's size and turned Ohio into the gateway for its great western expansion.

Z

Z is for Ohio's Zoos, too many to name them all. North, south, east and west...awesome animals, large and small.

THE TOLEDO ZOO

COLUMBUS ZOO AND AQUARIUM

ZOOMBEZi Bay

CLEVELAND METROPARKS ZOO
SECURING A FUTURE FOR WILDLIFE

akron ZOO

the Wilds

CINCINNATI ZOO & BOTANICAL GARDEN

Wolves wail in Cleveland? Tortoises tromp in Toledo? Alpacas ascend in Akron? It's true! Exotic animals aren't supposed to be able to live in Ohio's cold, but they do! The only catch is that you have to see them at one of Ohio's six world-class zoos. Thousands of animals from all over the world live here. So why not say "hi" to the hippos in Cincinnati? Or pose for a picture with the Columbus Zoo's polar bears? The choice is yours, but no matter what you decide, you can do it all without leaving the Buckeye State.

O is for Ohio, a state you just got to know.
Certainly a sweet one, "Hang on Sloopy. O - H - I - O!"

Made famous by Dayton's Rick Derringer, "Hang On Sloopy" is a song that's near and dear to all Buckeyes. Besides being the official state rock song, it's played at major sporting events throughout Ohio, including every OSU home football game. Fans watching at the world-famous Horseshoe chant, "O - H - I - O" while mimicking the shape of the letters with their arms.

CPSIA information can be obtained at www.ICGtesting.com
Printed in the USA
BVIW121149030719
552595BV00006B/40